Marjory's River of Grass

Marjory Stoneman Douglas, Fierce Protector of the Everglades

Written and illustrated by

Josie James

Christy Ottaviano Books

Little, Brown and Company

New York   Boston

This book is dedicated to the memory of Karsten Pohl, beloved husband and father, and our beautiful children, who showed me the Everglades.

"There are no other Everglades in the world. They are, they have always been, one of the unique regions of the earth, remote, never wholly known. Nothing anywhere else is like them."

—Marjory Stoneman Douglas

The wind rustled the saw grass. Softly, it sang as it stretched endless and free into the gathering clouds.

The dredgers roared as they tore into the peaceful landscape. The plan to transform this untamed wilderness into a human-made machine had begun. Each natural path the water chose to flow was destined to be straightened, forced into rigid canals, or locked behind dams.

Someone had to put an end to this foolishness.

Wearing her signature wide-brimmed hat and string of pearls, Marjory Stoneman Douglas persuaded the biggest and strongest politicians, government agencies, and land developers to protect and restore the largest subtropical wilderness in the United States—
the Florida Everglades.

How did Marjory become best friends with four thousand miles of wetlands?

That story began long ago,
in the year 1915.

The leaves were changing as Marjory boarded a train in Massachusetts. She was on her way to Florida to live with her father, Frank Stoneman. Marjory had not seen him since she was six years old.

In Miami, Marjory's clothes grew heavy in the heat. When her father arrived, he simply said, "Hello, Sweetheart," as if they had seen each other the day before. She noticed that they had the same eyes.

Marjory explained she had recently graduated from Wellesley College and that she wanted to be a writer. Her father was a writer, too. Frank worked for a newspaper called the *Miami Herald*.

"What do you write about?" she asked.

"The Everglades," Frank answered.

Marjory liked this new word and asked, "What is the Everglades?"

"It's South Florida's wetlands," Frank said. "Many consider the Everglades to be a useless swamp. The land becomes more useful when drained and sold to grow crops, build homes, and raise sugarcane."

"Drained?" Marjory questioned. "Father, how do you drain a swamp?"

"It's called reclamation," he explained. "The fresh water that drenches the Everglades is pushed into constructed canals, locks, and dams. Now when it rains, the drained land floods, and when it's hot, it burns and turns into dust."

Marjory was happy that she and her father had something in common. Draining the Everglades didn't sound like a good idea to her, either.

LANDS NOW BEING RECLAIMED
RICHEST SOIL IN THE WORLD.
DEEP MUCK, BIG CROPS, YEAR AROUND SEASON
BUY NOW AT GROUND FLOOR PRICES.
REALTY SECURITIES CORPORATION, Miami

Swaying palm trees replaced the towering maples of New England.
Marjory bought a bathing suit and swam in the warm, salty sea.
It felt very different from riding a bike through
the snow in her long, heavy skirt.

Marjory worked alongside her father as society editor at the *Miami Herald*. She proved to be an excellent journalist. Soon she had her own daily column.

As a freelance writer, Marjory explored life in the Everglades in depth. Her stories were published in national magazines like the *Saturday Evening Post* and reached thousands of readers.

Amid her success, Marjory had to face a sad truth: The Florida she loved was disappearing before her very eyes.

The land boom of the 1920s attracted people from all over the country who were hungry to gobble up reclaimed Everglades land. One such newcomer was Ernest Coe, a landscape architect and plant lover from Connecticut.

He confessed that he had come to Florida to get rich but, in the process, found something more valuable in the Everglades. With the assistance of his magic lantern slides, he lectured about the impending extinction of rare flora and fauna.

"Egrets, roseate spoonbills, and herons are pillaged for their feathers," Ernest explained. "Alligators are hunted for their skins. Orchids are plundered. Canals foolishly drain away the precious water that plants and animals need to survive. If Congress does not protect the Everglades in the form of a national park, this unique beauty will be lost forever."

Ernest's words cast a spell over Marjory.

The Everglades wilderness called.

Marjory joined Ernest's Tropic Everglades National Park Commission. There were no majestic mountains or towering waterfalls to impress the National Park Service, but Marjory and Ernest believed that the subtle wonders of the Everglades were just as spectacular.

Ernest's persistence paid off. Before long, a delegation from Washington, DC, was on its way to South Florida.

In January 1930, the members of the National Park Service arrived. Marjory and Ruth Bryan Owen, the first congresswoman in Florida, joined the men on a zeppelin ride to marvel at the hammocks, the mountains of the Everglades.

Birds sang, woodpeckers tapped, frogs croaked, and insects buzzed as the observers slogged beneath the majesty of a cypress dome.

The Everglades put on a spectacular show for the delegation. Roseate spoonbills displayed their bright pink feathers. The sunset lit the sky on fire, and more stars than the visitors had ever seen scattered themselves across the darkening heavens.

The National Park Service told Congress that the Everglades was a subtropical paradise full of the most fascinating life on land, in air, and in water, worthy of becoming a national park.

Unfortunately, Congress was in no rush to make a final decision. This meant trouble for Marjory, Ernest, and the Everglades.

Politicians, land developers, and sugar growers took advantage of the time. More wetlands were reclaimed, leaving little water and habitat for the plants and animals. Less water flowed south through the Everglades to Florida Bay. Salt water slowly began to seep into the Biscayne aquifer, the main source of drinking water for the people of South Florida.

Time was running out.

Marjory had to convince her fellow Floridians that the Everglades was not a useless swamp! But how?

Luckily, in 1940, a great idea leaped into her life like a tarpon.

24

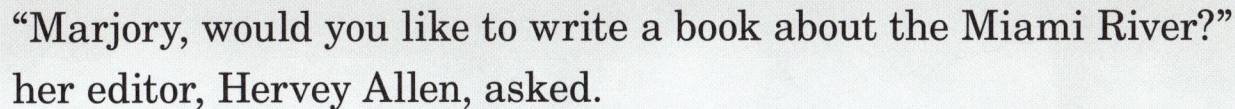

"Marjory, would you like to write a book about the Miami River?"
her editor, Hervey Allen, asked.

"How about I write a book about the Everglades instead?"
she boldly suggested.

"Prove that the Everglades is a river, and
the project is yours," he replied.

25

Water soaked her shoes as she spoke with Garald Parker.
He worked as a hydrologist, a scientist who studies water.

Marjory asked him the most important question of her life:
"Is the Everglades a river?"

Garald paused and said, "A river is a body of fresh water moving more in one direction than the other. The water of the Everglades moves just a few feet per hour. It travels about one hundred miles through the saw grass from Lake Okeechobee, west to Big Cypress, and south to Florida Bay.

"It *is* a river," the hydrologist most definitely declared.

The Everglades had a long story to tell. After many years of research and study, Marjory began her book with these words:

"The miracle of the light pours over the green and brown expanse of saw grass and of water, shining and slow-moving below, the grass and water that is the meaning and the central fact of the Everglades of Florida. It is a river of grass."

Readers were fascinated by a main character that was not a person but a vast subtropical wilderness. A different understanding of Florida's wetlands began to grow.

The subtitle of Marjory's book, *River of Grass*, soon replaced "useless swamp" to describe the Everglades.

In December 1947, Everglades National Park became the twenty-eighth national park in the United States.

"Here is land, tranquil in its quiet beauty, serving not as the source of water, but as the last receiver of it. To its natural abundance we owe the spectacular plant and animal life that distinguishes this place from all others in our country," declared President Harry S. Truman.

Marjory congratulated Ernest Coe on his remarkable achievement.

He was worried and said, "Congress approved only a fraction of the land I requested. Marjory, our fight to preserve the Greater Everglades has just begun and might never end."

Ernest was right. His park and Marjory's book were not enough.

Drainage continued. In the 1960s, the Army Corps of Engineers straightened each bend of the Kissimmee River, the headwaters of the Everglades, into a fifty-six-mile-long canal called C-38. This wetlands machine allowed even less water to flow down the River of Grass.

Then something even worse happened.

"Marjory, construction on the largest airport in the world has begun in Big Cypress," explained environmentalist Joe Browder. "If the project isn't stopped, we will be forced to say goodbye to the Everglades. It will become as dry as a desert."

"What is there left for me to do?" Marjory answered. "I am almost eighty years old."

"We need your voice," Joe retorted. "No one is going to argue with the woman who wrote *River of Grass*."

Joe suggested Marjory enlist the help of others to join in the fight. She named her organization Friends of the Everglades. Fifteen to twenty new members joined each time she addressed a gathering about the importance of protecting the Everglades. Marjory collected the one-dollar membership fee in her purse. Within a year, they had five hundred members. Soon there were three thousand in thirty-eight states.

With the help of the many Friends of the Everglades members, the construction of the airport ceased. On October 11, 1974, the land was designated Big Cypress National Preserve, the nation's first national preserve, as an act of Congress under President Gerald Ford,

"to assure the preservation, conservation, and protection of the natural, scenic, hydrologic, floral and faunal, and recreational values of the Big Cypress Watershed in the State of Florida."

Throughout the 1980s and 1990s, scientists discovered the Everglades to be one of the most diverse ecosystems on Earth and confirmed the area as an irreplaceable source of South Florida's fresh water. Despite these facts, the disregard for the environment continued. Marjory's spirit remained strong. She was often discouraged, but she never gave up.

On April 7, 1998, Marjory turned 108 years old. Her eyes could no longer see the Everglades' dazzling blue heights of space, nor could her ears hear the rustle of the thirsty saw grass prairie. It was time for her to rest.

The cycle of life continues. The dry and wet seasons come and go.

Florida's precious water still flows because new generations continue to protect and defend Marjory's dearest friend, the Everglades.

AUTHOR'S NOTE

Marjory Stoneman Douglas was born far away from the Everglades in Minneapolis, Minnesota, on April 7, 1890. Her activism began at Wellesley College, where she developed her writing and public speaking skills. These became Marjory's superpowers. She used them fearlessly. Marjory was an active participant in the suffrage movement and supported women's rights. During World War I, shortly after she moved to Florida, she enlisted in the United States Navy and became a yeoman, first class. A year later, she joined the American Red Cross and was stationed in Europe until January 1920.

Marjory skillfully used her daily column and her position as an editor at the *Miami Herald* to address the poverty and segregation that plagued Miami. Throughout her life, Marjory stayed true to the Quaker ideals of integrity, equality, and stewardship with which she grew up. Well into her ninth decade, she aided the Florida Rural Legal Services. She denounced the exploitation and substandard living conditions of migrant laborers, especially those employed by Florida's powerful sugar growers.

> "There is really nothing more invigorating and thrilling than to ride far out into the Everglades, or into the edge of them, and watch the clouds piling high in great soft toned mountains, over the wide expanse of lavish green earth, which opens out on every hand clear to the sharp cut rim of the horizon."
> —Marjory Stoneman Douglas
> The Galley, *Miami Herald*, June 23, 1923

> "Be depressed, discouraged, and disappointed at failure and the disheartening effects of ignorance, greed, corruption, and bad politics— but never give up."

During her lifetime, Marjory received many awards. In 1993, at age 103, she was awarded the Presidential Medal of Freedom. In 1997, 1.3 million acres, 86 percent of Everglades National Park (which had been set aside as wilderness in 1978), were named the Marjory Stoneman Douglas Wilderness to recognize her contributions to the protection of the environment.

In 1998, Marjory died peacefully at her home in Coconut Grove. Her ashes were scattered by Everglades National Park rangers across the Marjory Stoneman Douglas Wilderness. She was inducted posthumously into the National Women's Hall of Fame in 2000.

Her book, *The Everglades: River of Grass*—which the US Congress said "defined the Everglades for the people of the United States and the world"—remains a classic of American environmental literature.

Marjory Stoneman Douglas dressed as boldly as she spoke. Upon her visits to the Everglades, she famously donned an elegant dress, practical shoes, matching purse, wide-brimmed hat, and string of pearls. Marjory's fashion choices showed respect for the biological masterpiece she so fiercely represented. Yet Marjory understood that Floridians do not have to explore the Everglades to appreciate that the subtropical ecosystem in their backyard is the source of every drop of their precious water.

WHERE DOES YOUR WATER COME FROM?

In South Florida, drinking water comes from the Biscayne aquifer, an underground river replenished by the water that flows through the Everglades. This source supplies water to over four million people every day.

IS YOUR DRINKING WATER CLEAN?

Not all people in the world have access to safe drinking water. According to UNICEF, 1.8 billion people worldwide live without access to safe water. Each year over 800,000 people die from diseases directly caused by unclean drinking water and unsafe sanitation.

There is also unsafe drinking water in the United States. Toxic chemicals leak into the water system, and pipes made of lead, which can cause lead poisoning, are a public health problem. Also, groundwater near mines can be polluted by heavy metals like uranium.

DO YOU HAVE TO WALK MILES TO GET WATER?

People in the developing world walk an average of 3.5 miles a day to gather water. Some people in the United States have to buy bottled water or travel to get water because of drought and the contamination of water sources.

DO YOU USE WATER WISELY?

The average total home water use for each person in the US is about fifty gallons daily. The states of Arizona, California, Kansas, Nevada, New Mexico, Oklahoma, and Texas experience frequent water shortages.

WHAT CAN YOU DO TO LIMIT WATER WASTE?

- Take shorter showers.
- Wash your clothes only when they are dirty.
- Don't leave water running when you brush your teeth.
- Install a rain barrel for landscaping.
- Install water-saving appliances and faucets.
- Check your town's website for more ways to save water.

SELECTED BIBLIOGRAPHY

Davis, Jack E. *An Everglades Providence: Marjory Stoneman Douglas and the American Environmental Century*. Athens, GA: University of Georgia Press, 2009.

Douglas, Marjory Stoneman. *The Everglades: River of Grass*. Pathfinder Books, 2015.

Douglas, Marjory Stoneman. "A Tale of Two Women: Marjory Stoneman Douglas and Marjorie Harris Carr." Produced by Florida International University Learning Resources for FIU/FAU Joint Center. Filmed June 15, 1983, at the Douglas House in Coconut Grove, FL.

Douglas, Marjory Stoneman. *The Wide Brim: Early Poems and Ponderings of Marjory Stoneman Douglas*, edited by Jack E. Davis. Gainesville, FL: University Press of Florida, 2002.

Douglas, Marjory Stoneman, and John Rothchild. *Marjory Stoneman Douglas: Voice of the River*. Sarasota, FL: Pineapple Press, 1987.

SOURCE NOTES

3   **"There are no other Everglades":** Douglas, *The Everglades*, 3.

9   **"Hello, Sweetheart":** Douglas and Rothchild, *Marjory Stoneman Douglas*, 96.

27   **"A river is a body of fresh water moving":** Douglas and Rothchild, *Marjory Stoneman Douglas*, 191.

28   **"The miracle of the light pours":** Douglas, *The Everglades*, 3.

30   **"Here is land":** Everglades National Park Dedication Program, December 6, 1947.

33   **Fifteen to twenty new members:** Douglas and Rothchild, *Marjory Stoneman Douglas*, 226.

34   **"to assure the preservation":** "Everglades Jetport," National Park Service, last modified May 1, 2020, https://www.nps.gov/bicy /learn/historyculture/miami-jetport.htm.

38   **"There is really nothing more":** Douglas, *The Wide Brim*, 34.

38   **"Be depressed, discouraged":** Davis, *An Everglades Providence*, 529.

39   **According to UNICEF, 1.8 billion people worldwide live:** *Triple Threat: How Disease, Climate Risks, and Unsafe Water, Sanitation and Hygiene Create a Deadly Combination for Children* (New York: United Nations Children's Fund [UNICEF], 2023), https://www.unicef.org/media/137206/file/triple-threat -wash-EN.pdf.

39   **People in the developing world walk an average of 3.5 miles:** "For nearly a billion people, a glass of water means miles to walk," CNN, April 29, 2011, http://www.cnn.com/2011 /HEALTH/04/29/drinking.water/index.html.

With gratitude to Allie Hartmann of Friends of the Everglades for her review of this book.

About This Book

The illustrations for this book were created with Photoshop on a MacBook Pro, using Astropad, an iPad Pro, and an Apple Pencil to digitally paint. This book was edited by Christy Ottaviano and designed by Karina Granda. The production was supervised by Rina Guo, and the production editor was Jen Graham. The text was set in New Century Schoolbook, and the display type is Boucherie Flared.

VISIT

Everglades National Park and
the Marjory Stoneman Douglas
Wilderness
Ernest F. Coe Visitor Center
40001 State Road 9336
Homestead, FL 33034
(305) 242-7700

Shark Valley Visitor Center
36000 SW 8th Street
Miami, FL 33194
(305) 221-8776

Big Cypress National Preserve
33100 Tamiami Trail East
Ochopee, FL 34141
(239) 695-2000

Marjory Stoneman Douglas
Biscayne Nature Center
6767 Crandon Boulevard
Key Biscayne, FL 33149
(305) 361-6767

Florida Museum of Natural History
University of Florida Cultural Plaza
3215 Hull Road
Gainesville, FL 32611
(352) 846-2000

JOIN AND PROTECT

Friends of the Everglades
3727 SE Ocean Blvd, Suite 200,
Stuart, FL 34996

11767 South Dixie Highway,
Suite 232, Miami, FL 33156

(305) 669-0858

info@everglades.org